Radiesthesia IV
Geopathic Stress

ISBN: 9781787234499

Radiesthesia IV

Geopathic Stress

Insights, forms and solutions

by

Christopher Freeland Ph.D

CONTENTS

INTRODUCTION

There is one thing common to life – all life. It is consistently subject to natural forces. No matter how hard one tries to work with or against those energies, they have the final word. They share the same life force as we do, there is no fundamental difference, they are a living entity. One might even feel they have an intelligence. Indeed, they do, but that is not what science would have us believe, for if anyone who has studied and worked with water will tell you, that is the only honest conclusion one can reach. When one stops being a person, a scientist, a dowser, a compartmentalized individual, there arises a chance – I repeat, a chance – that one is. "Is" as in "I am". In our materialistic regime, for our entire time here on earth, that angle is reinforced, but rarely considered. In a more spiritual, wholistic regime, that is the angle to be examined and assimilated.

The natural conclusion is a complete unity, a perfect totality with everything in its place, in constant but impeccable flux. Thank goodness, we do not see this as it is, it would be far too frightening for our fragile constituencies.

That being so, once more we are in that zone where it is practically impossible to prove what is said. A place where us moderns with our sophisticated know-how have a little difficulty accepting that we are not in total control.

Can vulgar geophysical anomalies actually affect us?

More specifically, we know that stress – in all its forms – is, and always has been, a problem for humans, animals and plants, but how can the energetic influences rising from the earth be included in that category?

The ancients (Sumerians or Babylonians rather, Chinese, Egyptians, Greeks [normal they applied what they learned at school in Egypt] and Indians) knew not to build on zones that are affected by what is below ground and works its way up to wreak havoc on the surface. We know that fact by inference alone, because no ancient monument is built over such problem zones, but that is a strong argument when attention is paid to that important fact.

Unawareness, ignorance and greed make for poor bedfellows as we well know, however a very large majority of the urban population pay with their health as a result.

There is not much one can do about ignorance, if a person wishes to ignore what they know is bad for them, that is their choice so best to leave them be. Unawareness is another matter, and that is where this booklet will possibly be of use.

We have a wonderful capacity of perception as humans, and an even greater ability to find solutions once that awareness is put in motion, connected to our intuition and intelligence. Especially when suffering is involved.

WHAT IS GOING ON?

Any sensitive human or animal is aware of their surroundings. They 'feel' things, now while they may not be able to say exactly what it is they are feeling, they are sensible enough to remove themselves from its influence.

When one takes a wholistic view of a mammal and their constituent parts, it becomes apparent that there is an interface between the individual body and the external world. When I say wholistic in this context, I mean an overall, non-differentiated view, as much as that is feasible for a simple human to assume.

On one side there is the harmonious movement of energy, the vital force as expressed in the living organism organizing blood, breathing, the nervous system and the vital organs; all operating thanks to a constant flow. On the 'outside' is an apparently constantly evolving sea of energy, gross and subtle on our scale of scientific assessment.

There is a common denominator to the two, and that is movement.

Remember your physics classes? Who says movement or current, says magnetic field at 90° to the flow.

Just because it is a tiny force does not mean that it is innocuous. Far from it, it is bound to have an effect. On what, to what extent and how are the rub, but it IS going to have an effect.

In all probability outside of the laboratory, and even then, there are no instruments sufficiently refined to measure such sensitivity, those receptive individuals are more magnetically sensitive than others, and that is where I maintain the problem lies.

Low-field magnetic influences, so sorely neglected by just about every branch of science, have so much to account for. Going back in time and to China, we find not only recognition of the dangers of underground water, but what is far more pertinent, the movement of energy in the meridians, as recognised in traditional Chinese medicine.

This magnetic component is especially relevant to us earth-dwellers because those frequencies coming from the ground rise up to the surface where we hang out, and can work serious mischief.

It is nigh on impossible to determine all the effects at work in our environment. Let alone which ones are impacting us, because not only are our metabolisms different, but our magnetic fields are too. There are so many potential factors at play that it is positively daunting trying to calculate what is going on.

My work is simplified and made much easier by questioning with a pendulum. Now while that may not fit in with everyone's belief system, I know of no other method which provides answers in which I have total confidence, because experience over the years has proved it to be utterly trustworthy.

We know that the majority of medical diagnosis is symptom-based rather than frequency-oriented, and there is nothing wrong with that. What is wrong is that the remedy is systematically provided by purveyors of chemical compounds, irrespective of the initial observation and the observer's experience.

That is the way things are and there is little one can do about it, except go maverick and refuse the pill. Happily, more and more people seem to be doing that by searching for alternatives; and there are many valid options available.

Having said that, some people do pick up on what is going on around them, including underground, but not many know of the potential issues, and even fewer, who know that you can actually do something about it once you have discovered what it is.

It is a matter of sensing – one of our 365 senses I would add – because you are subject to the specific influence more than the next person as your particular metabolism works like that. However, unless you are familiar with Nature's ways, there is very little chance that you will associate what you are feeling with what is going on below ground.

Here is a photo of what the passage of underground water can do to trees when they find themselves on the surface above a subterranean flow of water.

Even if you see this kind of warp, it is unlikely that you will know what you are actually looking at, as no one teaches us these affairs. What is more concerning is that what you are looking at is harmful, for us were we to spend any length of time over such terrain.

We can move out of the way, the trees can't.

Here is another photo of the effect on trees taken in the botanical garden just outside of the town of Peradeniya in Sri Lanka.

Most of us are familiar with the effect of wind, especially on the coasts of wind-swept lands where the trees are stunted and deformed in the prevailing direction of the incoming force. Unlike that phenomenon, these trees are simply over an underground stream which works its way down to the nearby river, but on the way the magnetic field of the underground current, as it reaches the surface of the earth, is sufficiently strong to impact the frequencies of the living tree and warp it in varying directions – so not the influence of wind. A tree the size of the specimens above is probably conveying 3,000 litres of water up and down its trunk every day and night. In all probability, it is the magnetic influence of that flow which interferes with the original natural tendency of the tree to grow straight. For when Nature produces straight lines, it is for the effortless ease of flow.

Such influences cause deformation on a cellular level after a certain time – a completely unknown factor in the equation – and can be dangerous, even mortal for humans.

There is no need to call upon any so-called authoritative science in my mind, for the simple reason, there is none that studies the effect of geology on the inhabitants of the troposphere, or that would be able to either explain or remedy what is going on.

It is of interest, however, that the medical profession has since around 1900 recognized GS as a pertinent factor. A medical doctor in Germany, Gustav Freiherr von Pohl, found that a number of his cancer patients all came from the same village. When he asked a dowser to go and check, the latter found that all the patients were subject to geopathic stress (GS) in the form of an underground flow of water over which all of their homes were located along one side of the street, and in addition they were all sleeping in the zone of influence.

Since that time, there are a few doctors paying attention to this phenomenon, but far too few to make the necessary impact of awareness of the general public. The damage caused can only increase as a result of unawareness and irresponsible construction.

Any builder with a few years of experience under their belt will be aware of the effects of certain terrains, yet be unable to explain the causes of geological phenomena. Subsidence is one of the more easily explained problems. The capillary action of subterranean water can only be determined on a micro-scale with radiesthesia; the excellent quality of the British hydrological survey maps around the world are most useful but on a large scale. Rising damp, poor drainage and fungal issues can be happily, albeit expensively, resolved. Solutions are there for a restricted number of problems. GS can cause cracking in concrete or walls, often along the line of flow of the unseen influence, frequently leading to insect invasion or infestation – they love the low-energy areas.

We are, by and large, totally unaware of the forms and patterns assumed by energy in its movement. The exceptions, of course, being found in aerodynamics and certain disciplines of the physic sciences. These flows are inherent and essential to life, even if we do not appreciate that. Movement is the agent of change.

Those forces are the patterns of life, from which we can learn a great deal if we simply make the effort to observe, and even better, cautiously experiment with them. Conversely, we ignore them at our peril, for if we really want to understand how we can fit harmoniously into Nature, we need to learn Nature's ways rather than attempting to impose our ways on Nature.

TELLURIC FORCES

Telluric is the term used to define what comes from the earth – the Yin energies perhaps. These forces derive from a number of causes, such as geological formations, faults, underground streams, the crossing of waterways, underground caverns, mineral deposits, and so on.

An individual geological formation can, and generally does, generate an electric current with its companion magnetic field, which of course interacts with the surrounding environment, knocking on up to the surface where it is released from the constraints surrounding it below ground. That force apparently continues rising until such time that the incoming cosmic forces overcome and neutralise it, so to speak; which will happen at high altitude.

It is extremely hard if not impossible to determine the precise nature and origin of a combined field of force in its evolution, as not only is there a lack of available instrumentation to measure and gauge the energetic strength of that force, but it is further modified by additional and highly random factors, such as time of day, atmospheric conditions of temperature and pressure, planetary influences, and so on.

These are the telluric forces that can adversely affect humans and animals, which contribute and form one aspect of what we commonly refer to as geopathic stress (GS).

These geological events have their own footprint, so to speak, when the energy they produce reaches the surface of the earth and interacts with whatever is present there. As these events evolve, it naturally becomes quite difficult to appreciate and assess their effect because we do not recognize their impact, once again lacking the instruments to measure their effect, and even less, do we record statistics in their regard.

When general health problems (which can, of course, assume mental, emotional, or physiological dimensions) arise, however, we are keen to find causes, and if these energy forms are part of the equation, it might be easier not only to explain the source but, far more importantly, find a solution.

Of the most common telluric events, one finds:
- Underground water– in any form– flowing through any underground passage, pipe, fracture, or fissure. The resultant magnetic field can be surprisingly high in frequency and strong in effect. The field fluctuates depending on multiple factors, including what the water is conveying, the speed at which it flows, and/or whether it interacts with any other earth energy. The earth's natural energy field is particularly influenced by the crossing of two or more watercourses, or by the crossing of a watercourse and another type of energy line.

On a scale of worst to good, the crossing of two underground waterways, irrespective of their individual depths, is the most dangerous phenomenon for humans with regard to biological damage, due to the intensity of the combined magnetic frequencies on reaching the surface.

- Geological formations, with their different mineral and magnetic components, are responsible for a lot of mischief, but mainly as a function of their ability to allow the flow or directional thrust of water; deposits of magnetically or gas charged rock are also to be contended with.

- Geological faults, caused by movement in the tectonic plates, or the result of the upheavals the earth has experienced, also affect the energy field at the surface. While they are less common, mainly because they do not often present zones suitable for construction, they can be very harmful.

- Underground caverns cause problems for the inhabitants above due to the resonance developed by the substances collecting or passing through the cavern zone. Invisible from the earth's surface, these holes are constantly creating energetic movement due to pressure, interacting with surrounding formations of a chemical and/or physical nature, and the natural passage of the effluents upward to the zone of least resistance— where we reside.

Basically, these energies create frequencies that interact with the environment and the people who are in the vicinity at their point of exit. Obviously, but not necessarily so, those people with a weakened metabolism will be affected more than robust individuals.

The multiple categories of telluric forces as explained by many dowsers, although of interest in the way the earth possibly functions, are of no solid practical use here. They are one more anthropocentric (read money-grubbing) demonstration of knowledge because nothing of benefit can be derived for humanity, let alone Nature, even if you know that it is a "vile vortex", "Curry/Hartmann line" or some such at the root of the problem.

If you are of a practical rather than a hypothetical bent of nature, and are more interested in achieving concrete results— as it would appear our ancestors were— it is probably best to leave these investigations to the 'theoreticians' and focus on what can be really accomplished, irrespective of the fanciful, scientific names people give to phenomena that no one really understands, let alone prove, because they are essential components of Nature.

Remember that telluric forces are natural but constantly evolving influences; regrettably, we have little real understanding of them and perhaps even less concern as they stream into our environment, which is increasingly cluttered with vibrations emanating from power grids, electrical circuits, computers, mobile phones, and routers, amplified by metallic structures and furniture, all of which play havoc with our vibrational balance, leaving Nature's creation to struggle in their midst with very little help coming from the cosmic energy that in the past balanced the whole.

Ancient peoples such as the Celts and Egyptians, realizing the potential utility of these subterranean forces, integrated them into their vestiges and practices. Early modern humans in all probability used, among other devices, large standing stones— menhirs and dolmens— to offset the effects of these influences, or alternatively they steered clear of them and built their homes elsewhere. This type of geometric and natural "science" was probably transmitted at a later date by various orders— Templars, Cistercians, guild-members— because the effects of these forces are demonstrably active even today despite the major modifications made in recent times to many European cathedrals and monuments built during the Middle Ages.

As William Stirling in his 1897 book, *The Canon*, said "The priests were practically the masters of the world... freemasons, or some body corresponding to the mediaeval freemasons, with exclusive privileges and secrets required for building the temples under ecclesiastical authority, have always existed. And the knowledge we possess of the mediaeval freemasons is sufficient to show that their secrets were the secrets of religion, that is of mediaeval Christianity."

We have been left with no user manual as to how such arcane and exclusive processes operate, so enquiry and experimentation with a mind to re-creating something similar can lead to some fairly conclusive results. One of the first tasks to be accomplished is to locate telluric influences, especially the crossing of underground water, for that was a primary source of a very strong form of earth energy, which can be diverted and turned to benefit, as is the case with the system I use and explain below.

Philosophically speaking, it is one more demonstration of the remarkable way that Nature functions, showing that the existence of such harmful forms of energy can be altered, so providing a silver lining to our otherwise rather daunting worldview.

EFFECTS OF GS

The human body can reasonably be compared to an alchemical plant – transforming, metabolising, transporting and fulfilling its life purpose. Interacting throughout its life-time with the frequencies coming from external and internal forces, caught up in a permanent struggle for harmony – ideally enjoying health and happiness, despite the strange paradigm we find ourselves in modern times where so many forces are apparently at work to destroy us.

No matter the belief-system you opt for, there is every reason to accept that fields of force and the frequencies generated by them, play a substantial role in how we thrive or otherwise. When we are in good health, the body's ability to resist external influences, bacteria and viri, is strong.

Telluric forces disrupt frequencies where those forces exit the ground and combining with the environmental energies create what we now refer to as geopathic stress.

If those frequencies belong to you, especially where you are immobile for lengthy periods of time, e.g. sleeping or working at a desk, then without any doubt they will have an effect.

That effect is dependent on a range of recognized and unknown factors, so we have to maintain an open mind in this respect. Of those recognized factors are:

- The extent of the exit zone, which can range from a beam of 1-2 centimetres to areas of several tens of metres.

If that 1 cm beam passes through a human body, it can cause grievous harm in the zone where it transits.

Even if there is no body in the way, it will in all probability affect the environment and modify the ambient frequencies. That is the nature of the phenomenon, and one of the reasons why we have such diversity on this planet. Certain areas where we have benign vibes, and others where death reigns.

Animals, as a function of breed in general, either steer clear or thrive in these zones. Domesticated animals, dogs, horses, sheep, cattle and larger mammals don't like it. Cats, ants, spiders and creepy crawlies like it.

- The intensity and its impact on the exit zone; this would help explain the power spots found around the world. The complexity of such makes too large a subject to be dealt with here.

These telluric energies on reaching the surface encounter man-made frequencies and mayhem ensues, hence geopathic stress.

For most forms of life and its normal development on earth, a balance in natural energetic vibration is required, or for the human and animal, resonance in the nerve and endocrine systems. So, for every living cell, whether human, animal, plant or mineral to grow, develop, mature, age and finally die, from old age and wear rather than from imbalance, it must, mandatorily, vibrate harmoniously.

The effects of geopathic stress on humans can be slow or the contrary and they are generally cumulative over time. Even the World Health Organisation, if it can be taken seriously, reckons that 30% of buildings have what is known as Sick Building Syndrome, or structures subject to geopathic stress. I would put that closer to 60% from experience.

Factors affecting the frequencies as they reach the surface are multiple and remain to be catalogued and defined on a scale ranging from benefit to harmfulness. I say benefit because there are also what we know as 'power spots, as many of you know of, and have perhaps experienced. Call them what you will, there are zones where frequencies exiting the earth to the troposphere, which can be positively beneficial, although probably best not to spend too much time in their influence.

Even in dense urban environments, one finds these beneficial exit-points, and although they are not common, it would be quite feasible and relatively easy to enhance them to the greater benefit.

With regard to the more numerous harmful zones, life is further complicated (for humans at least) by factors such as metals (beds and chairs), chemicals, poisons & pollutants, malnutrition/diet, recreational drugs, alcohol, psycho-social stresses, immune system weakness, electrosmog, chemtrails and so on.

The nature of such problems is varied, but appears to cover a vast, if not the entire physical, emotional and mental range: e.g. addiction, ADHD, aggressiveness, allergies, anorexia, anxiety, arthritis, asthma, bed-wetting, bulimia, cancer, candidiasis, cell de-regulation and growth issues, chronic fatigue, cot death (SIDS), dementia, depression, diabetes, eczema, enzyme production, exhaustion, foetal development, food intolerance, glandular fever, headache, heart disease, hyperactivity, infertility, insomnia, intestinal disorders, lethargy, lymph problems, memory loss, miscarriage, multiple sclerosis, myalgic encephalomyelitis (ME), nervous disorders, obsession, pain, Parkinson's, premature birth, resistance to treatment, restless sleep, rheumatism, schizophrenia, sexual abuse, skin problems, stressed relationships in partnerships, stroke, tuberculosis, etc., etc.

The effects do not only concern health: houses with geopathic stress running through them are consistently slower to sell than those without (it would appear that people sense something's wrong even if they are not consciously aware of it); lightning also strikes such areas on a regular basis.

GS is often a factor in struggling businesses, accident black spots, failing technology, corruption, financial decay and bad luck in all its forms.

A real concern is that no matter what medicinal treatment a patient receives, whether allopathic, ayurvedic, homeopathic, TCM or naturopathic, the organism is, at a certain stage, incapable of assimilating the beneficial effects of medication caused by the adverse vibration that has built up in the body due to the magnetic field arising from the ground, and the ambient conditions.

An individual's sensitivity to frequency varies hugely, which readily explains one person's complete insensitivity compared with another's instant allergy.

The rather perverse approach of modern medicine, given that there cannot possibly be a pharmaceutical remedy for this condition, is to ignore it. So, we at least now have a choice and hopefully a better understanding of what is involved.

WHAT CAN BE DONE?

Initially for humans, GS often manifests as a sleep disorder which is perhaps explained by the fact that mitosis, cell replication, occurs during sleep. If a person sleeps in a GS-free zone, the cells develop in harmony with the original state of the cell. However, on the contrary, if the ambient surrounding frequency stops the cell from vibrating at the desired level, the cell becomes deformed in relation to the original. Imbalance and disharmony follow.

Bed and work are where most people spend the majority of their time. While it is sometimes sufficient to move the sleeping area or desk a few feet and the danger is avoided, that may not always be possible, or desirable.

Placing a copper or aluminium sheet under the mattress will reflect the radiating waves from water, for example when an underground stream passes below, but that will have a contrary effect if there is a geological fault, and its effect would be accentuated. The best solution is invariably to use a neutralizing system as it can further usefully offset the effects of the other components that complicate our lives: radio masts, high voltage power-lines, industrial processes, domestic electric appliances and the numerous sources of daily exposure to magnetic and electromagnetic fields.

They say that GS appears in diverse forms, such as lines, beams or combined pattern forms. That is quite possible and even probable; but it also varies according to the time of day and night, as well as a function of atmospheric conditions. Awareness of such variants seems quite academic to me, unless of practical use. I would like to provide solutions, not more fascinating information concerning the multifarious and perverse forms adopted by 'energy' as it interacts in our world, especially with the increase in technology of the last few decades and its devastating effect on our natural environment.

So, let's be practical and find what can be done.

Detection of GS is easy on humans. Using a pendulum, simply place the pendulum over the Triple Warmer 2 point (Gate of Humours) at the juncture of the little and ring fingers. If the person is not subject to GS, the pendulum will indicate a positive response, if there is GS, it indicates a negative response.

To my knowledge no scientific biomedical journals have published any articles in the English language concerning Geopathic Stress. Having said that, it would appear that some doctors take the notion seriously, so there is a glimmer of hope that some suffering may be curtailed, but there is no doubt whatsoever, if we find that it is an issue we have to deal with it ourselves to the best of our abilities.

The Chinese, in their practice of *feng shui*, have much to say about energies coming from the earth and the skies. *Feng shui* practitioners, however, keep their cards very close to their chests and do not share the secrets carefully guarded by the fraternity. While it is interesting to learn that the energy structure of Chinese cities has been compromised by the building of railroads and motorways, probably like cities everywhere else in the world, there is never any mention of practical solutions to remedy the unfortunate effects of disrupting energy flows. Most of these effects are constant and timeless in their duration, far outlasting our short life-spans, but I would question the pertinence of some of the solutions offered by *feng shui*, particularly if they have not incorporated the technical innovations of the past century in their computations.

The Chinese deflect energies thanks to a carefully crafted system developed over thousand of years. As I said before, the problem with this is that for them, the problem should not exist because the basic instruction, as found on a stone epigraph, is DO NOT BUILD YOUR HOUSE OVER UNDERGROUND WATER. We westerners did not hear about that and built any old how – and now pay the price.

I am not denigrating *feng shui* by any means, but as it is such an ancient science, it has seemingly not evolved with modernity, and the technologies we use today are far in excess of the frequencies found thousands of years ago. To meet those demands, we need more than colour, form and rearrangement. While the *feng shui* devices and applications modify energetic effects to a certain extent, they cannot overcome electromagnetic and/or nuclear radiation and major disruption of magnetic fields.

For that is what we are dealing with in a standard modern home of the western world in an urban environment:– electromagnetic radiation (the other radiation – the nuclear variety can be dealt with successfully with an orgone accumulator) and magnetic field disruption, due to the prevalent use of electricity, lighting, public transport, WIFI routers, mobile telephony and so on.

The sceptics among you will perhaps scoff at this. So be it, but remember that at no time – ever – was any research made to discover the effects, benign, adverse or indifferent, created by the use of electricity on the human or animal metabolism, nor on the natural environment.

It is too late now, so we have no option but to resort to what we can find to offset the manifest effects of our modern gadgets. We are not anywhere near ready to give them up, but that does not mean we have to resign ourselves to being their victims.

The choice is an individual one, because no one is going to tell you of the inherent dangers, even though the fatalities were commonplace in the development of radar for example. The mobile phone uses exactly the same microwave tech as the radar, plus a few more technologies in the device and scattered throughout the countryside, but that is not the subject here.

So, one does not have to renounce using the lights, the refrigerator, the TV or the car. But, please consider protecting yourselves, your families, animals, gardens and so on by altering some of the frequencies in your environment, for they are the culprits and the one area where we are able to intervene and provide some relief.

Apart from using one's common sense, for example: not keeping the mobile phone next to where you sleep, turning off the router at night, using a wired rather than Wifi connection, no bone conduction headphones, opt for an air tube headset, and so on.

A method successfully used and learned whilst living in Thailand involves using four round river rocks. The flow of the "magnetic" force from the rocks can be concentrated by working on the pattern of their layout, and the orientation, up or down, (negative or positive) of the rocks. This is useful if one has access to rounded rocks and sure that the stones will stay in place, because if one of them is removed, the whole collapses.

Incidentally, rounded rocks are systematically discovered in the foundations of European churches, unearthed during modern renovations (especially at Chartres Cathedral in the 1960s). Their magnetic charge is cause for surprise, especially when laid out in a harmonious geometric pattern.

The origins of another old remedy to the problem of adverse telluric energies have disappeared in the mists of time. It involves locating the source of the problem underground and planting an L-shaped iron or copper rod in the ground at the exit point as indicated with the rods or pendulum.

This method appears to be quite commonly used in western Europe, and I have frequently found it employed in the Irish countryside. It sometimes works well, but if the flow underground changes, as it does as a consequence of the seasons, drought and flooding, so does the effect.

This approach works at diverting the energy – like the Chinese method. But it is not an altruistic way of doing things. The energy is diverted elsewhere, to the neighbours, to the garden next door... ..

Similar in principle to the Roman Catholic method of exorcism, get rid of the problem out of your backyard, send the invading spirit elsewhere and we'll deal with the rest later!

I refuse to work like that, if at all possible. Assuming that if one can modify one of the components, it is possible, or at least this appears to be the case after fifteen or so years of using a method to offset the damage-causing telluric energy, to alter the whole equation and improve the final outcome.

Obviously, when considering the overall energetic picture in situations where one is requested to restore harmony, there are numerous factors to be taken into account. The combination of multiple influences— telluric, cosmic, environmental, human, unexplained historical or spirit forces— involves a fascinating process whereby the apparent complexity is best reduced to the minimum, with the removal/cancellation of one element often producing an improvement. Almost systematically, the one that is accessible, in a manner of speaking, is the telluric component. And it does spare the inhabitants having to sell the house and move.

THE ULTIMATE SOLUTION

We will probably never know how the ancient Egyptians reached the remarkable understanding and knowledge they demonstrated in their monuments, with their subtle capacities and phenomenal concealed know-how. We have, however, achieved a certain comprehension thanks to our modern (albeit primitive) measuring techniques. I say that because the precision cutting of the stones and the sarcophagus inside Cheops' pyramid cannot be repeated today with our 'sophisticated' cutting devices. Some of that know-how can now be copied and employed to the general benefit of one and all, if only we were prepared to experiment more.

That experimental understanding is the basis of what follows. It took me several years of searching for this information, and then several more experimenting, before adopting the system I propose today. Further details of the method can be found in my biography.

The idea is to incorporate two very precise measurements into the architecture of a building. This is what the Egyptians, Greeks, ancient Indians and Chinese were doing.

One of the measurements is a constant, namely 63.5 cms, an exact proportion of the distance from the centre of the earth to the poles – 6,350 kilometres; the other measurement is dependent on the exact geographic location of the site. That will remain a professional secret because it took me a great many years to discover it, a lot of intense studying and I suspect, a certain form of spiritual integration which ought to be shared in mutual respect, sincerity and appreciation before being revealed.

The most convenient method where I live is using eight pieces of wooden curtain rail or dowel, with a diameter of ½ or ¾ of an inch, that are then installed upright in your home, barn, property. The advantage of this system is that you do not have to rebuild, just install the eight pieces of wood discretely – behind a cupboard, in a radiator, attached with double-sided tape if needed.

The good news is that the effect of this is far greater in extent than the other methods mentioned above. I have seen in a town, an area of several kilometres covered by the neutralising effect. It all depends on when the harmonising effect is countered by another wall of energy rising from the ground as a result of another telluric force exiting the earth.

The even better news, however, is that once these rods are installed, the frequency of the area is instantly modified. On my scale – which might correspond to gauss or micro-Tesla – the measurement rises to a reading that also involves a frequency of a colour that the French radiesthesists, André de Belizal and Léon Chaumery, reported as white: a colour indicating calm and harmony. A frequency that does not correspond to the energetic forms that encourage negative energies, that some refer to as ghosts or spirits.

I am quite happy to calculate the two lengths for you and either install them myself, or send you the two measurements so you can have them cut and tell you where to install them, in exchange for modest remuneration. All I need is your address – postal, Eir or Zip code, to determine if you are over geopathic stress and we can then work something out.

The ground is now clear for the establishment of a harmonious frequency, which is almost palpable because the body's biomagnetic field is no longer subject to the aggression of the telluric influence. This same method used in the past using sacred geometry, geometric form, length, and proportion, can be reproduced to enable a resonance of frequencies conducive to equilibrium of both the environment and everything in it.

About the author

The basis of this text is founded essentially on a combination of practical experience, the study of theory from selected books, substantial personal practice and experimentation in a variety of methods and traditions. It all started in India in the early seventies when I studied Sanskrit so as to read the ancient texts in the original rather than relying on translation. In that phase, I had the opportunity to study with Swami Pranav Tirtha, a *dashnami sannyasin*, who initiated me into the Vedanta philosophy. Whilst with him I read, studied and assimilated the orthodox teachings of the Upanisads, the Brahma Sutra, Gita and multiple metaphysical and sundry texts of Hindu literature. I was ordained as a monk, with the name of Swami Chidananda Tirtha in May 1973. This period also furnished the occasion to study medicine with Dr Himatlal Trivedi, an Ayurvedic practitioner from Palitana, whom I accompanied in India and Africa in his practice amongst English and Gujerati-speaking patients. That involved study of the Hindu medical classics (Caraka Samhita, Sushruta Samhita, Ashtanga Hrdaya), with considerable practical experimentation of fasting and dietary regime on myself. Observation with Himatlal's guidance and explanation gave me a reasonable understanding of this medical art form.

Whilst living in France, I had the opportunity to study for a year with a French acupuncturist, who was persuaded to come out of retirement to teach Traditional Chinese Medicine (TCM) again. That grounding was then followed by many years of studying the Chinese classics: the Lingshu, the Huainanzi, the Suwen, along with in-depth reading of Soulié de Morant, Claude Larre and Elisabeth Rochat de la Vallée, in addition to extensive practice of moxibustion and acupressure. Whilst living in Chiang Mai, the opportunity arose to learn and practice a form of bio-energetics which involved a lot of practical moxibustion. The outcome of my studies of TCM affords a certain ease with this very complete approach to the human condition.

A vast amount of research, reading and experimentation with a very broad spectrum of subjects, traditions and cultures, combined with travelling and living among natives of other lands, along with my professional activity as a technical translator, specialized in nuclear and telecom technologies, for some twenty-five years in France have hopefully been turned to good advantage.

Study, practice and research into magnetism, laying on of hands, geomancy, radiesthesia and radionics add to my wholistic comprehension of life. A practice of organic farming combined with animal husbandry, special care for water and its supply have also led to my current understanding. I work more with the intention of clarifying what I think seems to be happening in the dimensions we evolve in, rather than any kind of dogmatic laying down of law.

Whilst I sincerely believe my opinions to be correct because corroborated by the pendulum and experience, it would be a substantial error to think this is the last word because it concerns uniquely what falls within my own sensory (all three hundred and sixty-five) parameters. Too much is changing too fast for our perceptual ability to stay abreast of events, even were we able to comprehend and adapt. Perhaps it is not for us humans to determine how the ordered immensity of Nature works; that would be most presumptuous and dishonest, but it does seem to be worth trying to establish a mode of operation that might serve as a guide or possible reference fitting into our journey through this phenomenal existence.

I stand on the shoulders, hopefully in rectitude and fidelity to the thrust of the original argument of many researchers, practitioners and remarkable people in aligning these words on paper. The words will, as always, be symbols of the generosity of Nature as she carefully keeps everything in its structured place, although the human component must be the most unruly, hence the hard lessons we have to learn if we aspire to some other form of existence than the purely material.

www.ingramcontent.com/pod-product-compliance
Lightning Source LLC
Chambersburg PA
CBHW080633030426
42336CB00018B/3186